LUFTWAFFE

Fueling up a Do 17 Z using an Opel airfield tanker truck.

AIRFIELD EQUIPMENT

Joachim Dressel/Manfred Griehl

Schiffer Military/Aviation History
Atglen, PA

An MG 08/15 with drum magazine for defense against low-flying aircraft in World War I.

Sources
— Archival writer, test results of the Ar 234 and Me 262
— Archival writer, LDv 16 and LDv 90
— W.J. Spielberger, *Die Rad- und Vollkettenzugmaschinen*
— W.J. Spielberger, *Die Motorisierung der deutschen Reichswehr*
— W. Oswald, *Kraftfahrzeuge und Panzer der Reichswehr, Wehrmacht und Bundeswehr*
— *Waffenarsenal* volumes 82, 94, 107 and 123
— K.H. Völker, *Dokumente zur Geschichte der deutschen Luftwaffe*
— K.H. Völker, *Die deutsche Luftwaffe 1933-1939*

On the front cover:
A Renault UE tractor, Sd.Kfz 630 (f), with an SC 1000 in front of a Ju 88 A5.

Photographic credits

AEG, 2	I. Mayer, 1
Bundesarchiv, 15	T. Mayer, 1
U. Balke,	MBB, 2
E. Creek, 3	J. Menke, 2
Deutsche Lufthansa AG, 6	H.J. Nowarra, 3
M.M. Dressel, 3	Opel AG, 3
M. Griehl, 6	P. Petrick , 4
Flughafen AG, 3	W. Radinger, 2
K. Francella, 3	U. Rohrbach, 1
P. Heck, 2	G. Schlaug, 10
Heinkel GmbH, 6	H. Schreiber, 1
Henschel, 1	Ph. Schreiber, 1
K. Kössler, 1	H.H. Stapfer, 1
R.P. Lutz, 1	B. Wittigeyer, 1

Translated from the German by James C. Cable.

Printed in the United States of America.
ISBN: 0-88740-482-0

This title was originally published under the title, *Fahrzeuge und Geräte auf Flugplätzen der deutschen Luftwaffe vor 1945,* by Podzun-Pallas Verlag, Friedberg.

Published by Schiffer Publishing, Ltd.
77 Lower Valley Road
Atglen, PA 19310
Please write for a free catalog.
This book may be purchased from the publisher.
Please include $2.95 postage.
Try your bookstore first.

We are interested in hearing from authors with book ideas on related subjects.

Vehicles & Equipment Found on Luftwaffe Airfields to 1945

Before the outbreak of World War I, branches of the military rarely concerned themselves with the theme of utilizing vehicles in combat. The conservative army leadership did not recognize the possibility of achieving great tactical advantages by motorizing its troops. Individual branches of the military were equipped with only a small number of trucks for transportation purposes. The requirement for additional vehicles during wartime was to be filled by requisitioning privately-owned civilian vehicles. It was abundantly evident when war finally did break out that this plan was insufficient. Production of vehicles started sluggishly due to a lack of sufficient personnel. As a result, there were 40,000 vehicles in the entire Kaiser's army up to the end of the war, 25,000 of which were trucks, while the French had 40,000 vehicles on the western front alone. The United States had about 100,000 vehicles at its disposal.

Application of vehicles in the *Fliegerkorps* came about slowly. Beginning in 1909 there were a few models of the Krupp-Daimler platform vehicle and Ehrhardt vehicles which were either anti-aircraft or anti-balloon capable, for example armed with either the 6.5cm **BAK L/35** or 7.7cm **K-Flak L/27 C7/1911**. Some of these vehicles were even lightly armored and cross-country capable.

Four-wheel drive was hardly utilized by army vehicles because its construction was difficult to assemble.

At the end of the war, the firms of Daimler-Benz and Krupp developed, as a replacement for the platform truck, the KD-1 as a carrier for the 7.7cm Flak and as a towing vehicle for the 15cm long-barrelled gun.

Further utilization of trucks was seen with the aerial spotlight platoons of the air defense units, most notably the truck produced by the MAG firm, which were also utilized in aircraft recovery operations.

A number of trucks served the *Geschwader* by transporting personnel, disassembled aircraft, canvas hangars and all spare parts. The urgent transport of aircraft by truck was of particular importance at the beginning of the war because aircraft ranges were still quite small and transfer flights to new deployment areas could not always reach full-service airfields. The aircraft were transported primarily over roads or by rail to the airfield of assignment. Even before the beginning of the war, military authorities had required "The aircraft must be quickly dismantled and able to be easily transported on roads."

The emergence of motorizing aircraft recovery operations in 1915 on the Western Front.

A horse-drawn AEG electrical generator which was used by many aircraft repair facilities.

Using a light *Wanderer* vehicle to transport an AEG-monoplane.

A three-ton army truck with a folding platform which was in the inventory of air forces stationed near Cologne.

Up to the beginning of the 1930s, motorizing the *Reichswehr* had to remain limited to army units due to limitations imposed by the Versailles Treaty. It was not until later that commercial tankers, fire, medical vehicles and trucks were used for transport and delivery tasks on the civilian airports. Because military aircraft were often stationed on these same airfields, they were included in the maintenance. At the beginning of the war, the Luftwaffe commandeered the majority of these vehicles. Additionally, there were a few vehicles specially constructed for the Luftwaffe:

Kfz.-No.	Use	Chassis
Kfz. 301	Antenna mast truck	med. x/c truck, Krupp L 3 H 163 Henschel/Magirus 33 G 1
Kfz. 302	light radio truck	light x/c truck, Mercedes G 3a Magirus M 206 Krupp L 2 H 143
Kfz. 303	radio intercept and light x/c truck	similar to direction finding Kfz. 302
Kfz. 305	med. truck with med. x/c truck	Opel Blitz standard trunk 3 t, Mercedes-Benz L 3000 S
Kfz. 309	hose tender	med. truck (c), Mercedes-Benz L 3000 S
Kfz. 317	Oxygen tanker truck	heavy truck (c), Krupp 5t.
Kfz. 323	Tanker/Pumper (fire)	med. truck
Kfz. 343		Krupp L 3 H 163, Opel Blitz 3t HS 33 D1
Kfz. 344	hose tender	med. truck (c) like Kfz. 309
Kfz. 345	fire truck	light truck (c), Opel Blitz 1.5t
Kfz. 346	fire hose truck	heavy truck (c), Krupp 5t
Kfz. 353	runway illumination	med. x/c truck, Krupp L 3 H truck 163, Opel Blitz 3t
Kfz. 354	instrument	med x/c truck, Krupp L 3 H navigation truck 163, Opel Blitz 3 t
Kfz. 384	aviation fuel	med. x/c truck, Opel Blitz tanker truck 3 t.

med. =* medium
x/c =* cross-country capable

Aircraft participating in the *Deutschland Flüge* were refueled from two-axle commercially available fuel trailers.

Above: Air traffic control facility with an attached hangar in Magdeburg-Süd. In the foreground is a fire tanker truck belonging to the airfield fire department.

Below: A typical barracks found on many airfields. The aircraft with tail number "D-AMAB" is one of the few four-engine He 116's.

Above: Location of the establishment of *Fliegerstaffel "Schlagter"*, from which JG 26 was formed, on their well-renovated airbase near Cologne.

Below: Camouflaged aircraft hangar on the Oranienburg Heinkel factory airfield during winter flight operations.

The building up of military airfields was already in advanced planning stages in 1934. Up to the middle of the 1930s, a total of 19 air bases were built up or newly constructed for use by the military. They were located in Braunschweig, Celle, Cottbus, Döberitz, Fassberg, Gotha, Grossenhain, Heldesheim, Jüterbog, Kitzingen, Lechfeld, Neuhausen near Königsberg, Neuruppin, Prenzlau, Rechlin, Schleissheim, Staaken near Berlin and Tutow. By 1937, Brandis, Erfurt, Gablingen, Liegnitz, Neubrandengurg, Schwäbisch Hall, Seerappen near Königsberg Stendal and Wunsdorf were added to this list.

Additionally, sea plane bases located at Bug on Rügen Island, Holtenau near Kiel, List on Sylt, Norderney, Pillau, Swinemünde Travermünde, Warnemünde as well as Wilhelmshaven were expanded and built-up. Due to the fact that not all of the peacetime airfields were suitable for use in wartime, the Luftwaffe leadership started with the build-up of so-called deployment airfields (*E-Häfen*). A portion of these airfields were constructed as peace-time fields due to a shortage of such fields, and new *E-Häfen* were built thereby establishing an extensive network of military airfields on the territory of the Third Reich by 1939.

The *E-Häfen* of the first ranking had road and railroad connections, underground fuel storage, equipment storage together with repair facilities. The take-off and landing area was almost square in shape. Landing had to be possible in poor visibility and at night. *E-Häfen* airfields of the second ranking had no such facilities. Provisions were made, however, for a 1000 meter long and 500 meter wide runway oriented to the direction of the prevailing winds, and a second runway perpendicular to it, either stretching across the main runway or simply connected on one end, forming a "T." Additionally, the airfield was to be suitable for instrument flying conditions. A connection with a road was seen as absolutely indispensable. This was intended to permit the aircraft to land and taxi onto the road where they would be resupplied with fuel and ammunition.

These *E-Häfen* fields were only to be utilized in the case of mobilization and were to remain otherwise camouflaged. To this end, terrain was sought out which was naturally suited to an airfield, or that which could be made suitable by removing a few obstacles in the shortest amount of time. Another provision was the "utilization as pasture land" wherein the inconspicuous preparation of airfields was possible. These areas were managed and utilized by civil servants (so-called "*Platzlanddienste*") who were to vacate the land in the case of mobilization.

Construction of the necessary buildings, those to be use in peace and wartime alike, was specified in an order dated 14 August 1937 from the *Reichsminister der Luftfahrt* and the *Oberbefehlshaber der Luftwaffe*. The first thing to be done in accordance with this was the erection of a barn, which was to house not only necessary airfield equipment but five dismantled barracks. Then came the construction of a house in a style which fit the local architecture and conditions, heavy or half-timber style with two stories and a living area of 65 square meters. Additionally, provisions were made for a large equipment shed with a room for an agricultural tractor, forge and oil storage, a room for agricultural equipment, and a room for fertilizer. Sizes of these rooms were standardized. A barn with an oxen stall and stone floor could be constructed if needed, which could be small or large depending on needs. An open barn (field barn) with a threshing floor could also constructed. The open barn occupied a 450 square meter area.

One of many captured Soviet tractors served to grate the runway surface on this airfield belonging to KG 54.

Above: Under a thin mesh camouflage netting, a Luftwaffe construction team lays down steel matting in a parking area fortification. One method which is still in use today.

Below: At the beginning of the assault on the USSR, a large number of construction vehicles were captured, along with tanks and trucks, which were employed in airfield improvements in the Hinterland. The Luftwaffe runway repair crews of today's Bundeswehr maintain a motor pool of special vehicles used to repair air base runways after an enemy attack.

In accordance with orders from the *Oberbefehlshaber der Luftwaffe* issued on 18 December 1935, the systematic expansion of the flight ground services as well as the creation of air district commands, air base commands and supply stations began in early 1936.

As members of the command authority for ground organizations and supply facilities, air district headquarters were created which received territorially limited command and control authority. The command point of a fully developed military airfield was the *Fliegerhorstkommandantur*, or air base headquarters. The commander of the unit stationed on that base, usually a Group Staff with three *Staffeln*, possibly including a *Geschwaderstab* and an additional *Gruppe* or a flight school, was traditionally the commander of the air base itself.

In order to relieve the air district headquarters in case of mobility, provisions were made to combine command locations for several air bases, deployment airfields and forward area airstrips. Several of these airfield districts, in turn, made up a new air district command. Prior to the outbreak of the war, the headquarters of the airfield districts were budgeted for because operation of the ever more extensive flight ground services required this intermediate level of control even in peacetime.

The following air districts and air district command centers came into being in the homeland prior to 1936:

			Air District
			2 Stettin
Air District Command I	Königsberg		3 Hamburg
Air District Command II	Berlin		4 Berlin
			5 Magdeburg
Air District Command III	Dresden:		6 Breslau
			7 Dresden
Air District Command IV	Münster:		9 Hannover
			10 Münster
			11 Giessen
Air District Command V	Munich:		13 Nürnberg
			14 Munich
			15 Stuttgart
Air District Command VI	(sea)	Kiel	

Bivouac area of the 7th Staffel of TG 1, with an MG 15 machine gun on a tripod for air defense of the Gruppe.

Prior to the start of the war, 64 Air Traffic Control bases, 133 air and sea bases as well as 19 special airfields had been established. The air traffic control bases served as command posts for staff elements of deployed *Geschwader*, and in peacetime were to assume the supervision of the airfields and their infrastructure as well as the reconnaissance of new forward area airfields and the construction of air strips and mock airfields. They were additionally responsible for preparing all airports for immediate use in case of mobilization. During wartime, the deployment of the airport engineering staff and troops fell under the direction of the Air District Commands. They were also to ensure the integration of the aircraft warning and air defense warning services, the establishment of communications and were to supply and equip those airfields located within their areas of responsibility. In March 1939, these airfields were occupied with three reconnaissance *Geschwader (H)* with 21 *Staffeln*, 13 reconnaissance *Gruppen (F)*, five Fighter *Geschwadern* with 15 Fighter *Gruppen*, a Destroyer *Geschwader* with three *Gruppen*, 11 Bomber *Geschwadern* with 30 Bomber *Gruppen*, one Stuka *Geschwader* with three Stuka *Gruppen* and six independent Stuka *Gruppen*, five Coastal Patrol *Gruppen* with 13 *Staffeln* as well as a Transport *Geschwader*

with two Transport *Gruppen* and a tenth *Staffel*.

Supply services included/encompassed in July 1939 ten aircraft *Gruppen* and a seaplane *Gruppe* thirteen air parks, eight main ammunition depots, 39 ammunition depots, and two sea ammunition depots, thirteen aircraft fuel tank facilities, 14 vehicle maintenance platoons, four aircraft arsenals, one seaplane arsenal, one air intelligence branch, five medium fuel columns, as well as ten motorized and thirteen train-borne aircraft parts suppliers.

In wartime, the command structure was altered in accordance with the conquered territory. Afterwards, the *Luftflotten* and *Fliegerkorps* looked like this:

Peacetime:	Wartime:
Luftflotte	
1: Berlin	Northern Russia
2: Münster	Belgium, Central Russia, Italy
3: Munich	France
4: Vienna	Southern Russia
5: —	Norway beginning 12 April 1940
6: —	Central Russia beginning 11 May 1943, later in the territory of the Reich

Fliegerkorps		Later:
I	West	Northern Russia
II	West	Central Russia, Sicily
IV	West	Southern Russia
V	West	Russia
VIII	West	Russia
IX	England	
X	Norway	Sicily, Greece
XI	Norway	West, Crete, Russia, Italy, Africa
XII	Holland	Belgium, Reich-territory

The aviation ground services were also altered during wartime. In addition to changes in the command structure, there were far-reaching deployments of vehicles outside the built-up airports and airfields. Commandeered vehicles came into use as field-expedient alterations were made to outfit the flying units and their maintenance services. Describing the numerous vehicles in use by the Luftwaffe would burst the seams of this publication. Only the most important vehicles and equipment can be described herein.

In the Fall of 1944, this air traffic control facility was formed with a closed-cabin Phänomen Granit 1500S dug in to protect it against enemy fire.

Camping trailers often served as mobile quarters for aircrews. This belonged to Hauptmann Tietzen, who was shot down on August 18th, 1940.

The tires of this Opel-Blitz truck are being protected against the sun with sacks and towels.

Even former theatrical wagons found their way into the Luftwaffe. This one was used as a "sleeping car" for a Kampf-geschwader.

In addition to specially-constructed vehicles, which received a nomenclature between 300 and 399, Wehrmacht vehicles were also in use by the Luftwaffe. On improved airfields, where there was no need for cross-country vehicles, commercial trucks from the civilian sector were very often used.

The Mercedes Benz LG 3000 together with the Henschel 33 D 1 and the Krupp L 3 H 163 was the most commonly used triple-axle fully cross-country capable vehicle. Almost 7500 of these vehicles were built between 1934 and 1937 and were mostly utilized in engineer and Luftwaffe units. In addition to its service as an aviation fuel tanker vehicle, and when fitted with a hut, it also saw use as a radio, antenna and radio direction finding vehicle, as did the Henschel 33 D 1.

The Krupp L 3 H 163 was in production from 1936 to 1938 and was the successor to the L 3 H 63, which was produced beginning in 1928 and had a 66kW normally-carbureted engine. With a hut mounted atop, this triple-axle vehicle did service as a radio intercept, radio, weather and printing press vehicle, among other things. In its reconnaissance units, the Luftwaffe utilized the Krupp vehicle as a photo vehicle (Kfz.353). At airfields it could often be seen in its role as a runway landing light vehicle with a two-man crew and a trailerable runway spotlight made by the Siemens firm. For its utilization as an oxygen tanker vehicle (Kfz.317), the Luftwaffe often utilized the commercially available 5-ton Krupp vehicle, which was powered by a six-cylinder normally-carbureted M 12 engine capable of 74 kW output. The oxygen produced for aircraft high-altitude systems produced under field conditions was subsequently placed in oxygen bottles. This system was permanently installed at almost every airport on the territory of the Third Reich.

The Krupp-Protze L 2 H 143 was primarily utilized by Luftwaffe units as a crew vehicle (Kfz.69) and as an air defense vehicle (Kfz.81). In anti-aircraft units, this vehicle served as a light *Flakwagen* (Kfz.81) with the 2-cm Flak 30 anti-aircraft guns as well as a light spotlight carrier (Kfz.83).

The vehicles of the three-ton class, like the Opel Blitz and Ford V 3000 were available as platform trucks for troop and transport purposes. The Luftwaffe only had a limited number of the cross-country version of this vehicle. There were additional versions of both chassis available with special-purpose superstructures.

The Wehrmacht leadership was occupied with the creation of a hut type superstructure for all three-ton trucks within the framework of the *Schell* program. The following firms took part in the construction of this chassis beginning in 1938, which intended for civilian use without all-wheel drive, and for military use as a four-wheel drive. When the war began, those vehicles in civilian service were commandeered and partly transformed to four-wheel drive. This was an attempt by the Wehrmacht to avoid the expensive costs of special construction programs.

Manufacturer's of Three-ton Trucks Within the Scope of the Schell-Program, 1938-1945:

Firm	Production Plant
Adam Opel AG	Brandenburg
Daimler-Benz AG	Mannheim, Gaggenau
Klöckner-Humboldt-Deutz	Magirus in Ulm
Ford	Cologne
C.F.W. Borgward	Bremen

This cross-country capable Henschel Typ 33 D1 with 3000kg carrying capacity was one of the most widely used vehicles in the Luftwaffe.

Trucks with special huts which were in service with the Luftwaffe received nomenclatures of Kfz.-Nr. 3051/1 to 305/137. The number after the slant bar is the code showing intended use and contents. These were only bolted on and could be easily removed for other transport missions. A list from July of 1943 shows the possible applications and gives an idea of the multi-faceted missions of the ground support troops.

All of the above-named three-ton vehicles served as chassis, but these huts could also be mounted on commandeered vehicles or trucks originally intended for other purposes.

Kfz.305 Hut Contents Variation

305/1	**Radio Teletype vehicle (KW)**
305/2	**Radio Teletype vehicle G**
305/3	**Radio Teletype Retransmitter**
305/4	**Voice Frequency Telegraphy vehicle**
305/10	**Amplifier vehicle**
305/11	**High power vehicle**
305/15	**Transmitter vehicle**
305/16	radio vehicle
305/17	radio vehicle
305/18	radio vehicle
305/19	radio vehicle
305/20	radio vehicle
305/21	radio calibration vehicle
305/22	radio direction finding vehicle
305/23	radio receiver vehicle
305/25	antenna tower vehicle
305/26	radio sonde vehicle
305/27	radio sonde direction finding vehicle
305/29	localizer beacon identifier vehicle
305/30	radio marker beacon vehicle
305/32	light beacon vehicle
305/33	light beacon vehicle
305/34	microwave communications vehicle
305/35	microwave communications vehicle
305/36	microwave communications generator vehicle
305/37	microwave communications generator vehicle
305/38	antenna vehicle
305/40	microwave communications calibration vehicle
305/41	equipment testing vehicle
304/42	flight radio testing vehicle
305/43	anti-interference (radio) vehicle
305/44	electronics interference tracking vehicle
305/60	avionics test vehicle
305/61	airframe workshop vehicle I
305/62	airframe workshop vehicle II
305/63	engine maintenance vehicle
305/64	avionics workshop vehicle
305/65	weather vehicle A
305/66	weather vehicle B
305/67	engine workshop vehicle I
305/68	engine workshop vehicle II
305/69	steering control test vehicle
305/70	parachute storage vehicle
305/73	command post vehicle
305/74	orderly room vehicle

305/74	clothing storage vehicle
305/75	medical equipment vehicle
305/77	sutler (canteen) vehicle
305/78	electric kitchen vehicle
305/79	orderly room vehicle
305/80	weapons maintenance workshop vehicle
305/83	vehicular maintenance workshop vehicle
305/84	vehicular spare parts storage vehicle
305/85	test vehicle for hydraulic and oxygen supply system
305/86	dental clinic vehicle
305/87	dental surgery vehicle
305/88	x-ray screen vehicle
305/89	x-ray screen and darkroom vehicle
305/90	photo vehicle I
305/91	photo vehicle II
305/93	water treatment vehicle I
305/94	water treatment vehicle II
305/95	water treatment vehicle III
305/96	water desalination vehicle
305/98	oxygen depot vehicle
305/99	oxygen bottle filling vehicle
305/100	navigation vehicle
305/101	navigation instrument measurement vehicle
305/103	localizer test vehicle
305/104	KTZ storage vehicle
305/105	fitter and pipefitter workshop
305/106	fitter and hydraulic workshop
305/107	machine shop facility
305/108	weapons and electronics workshop
305/109	mechanical workshop vehicle I
305/110	mechanical workshop vehicle II
305/111	workshop for fitters and welders
305/112	pipefitter workshop vehicle I
305/113	pipefitter workshop vehicle II
305/114	workshop vehicle for hydraulics and fitters
305/115	workshop vehicle for joiners, painters, and saddle
305/117	welding and forging workshop
305/118	vehicular workshop vehicle for engines, chassis and electrical systems
305/119	vehicle parts storage
305/120	crew vehicle for depot services
305/121	depot services equipment storage vehicle
305/122	decontamination vehicle
305/123	decontamination equipment storage vehicle
305/123	air compressor vehicle
305/125	flight service vehicle
305/126	starter vehicle
305/128	fueling hose vehicle
305/130	workshop vehicle for jettisonable munitions
305/131	workshop vehicle for aircraft guns
305/132	weapons workshop vehicle
305/135	workshop vehicle/mechanical workshop
305/136	engine workshop vehicle
305/137	workshop vehicle for aircraft fitters and armorer (TMZ)

For use in aircraft recovery operations and for the installation/removal of aircraft engines, a crane was mounted on the Borgward B 3000 chassis, and a rotating crane (Kfz.100) on cross-country capable truck chassis of the 4.5 ton class which also had towing gear, for example the MAN ML 4500, Büssing-NAG 4500 or the Mercedes Benz L 4500. The Opel "Blitz" also saw use as a crane in somewhat smaller numbers.

All of the usual vehicles in service with the Wehrmacht, particularly the "Einheits-PKW" (unit car), saw use as passenger automobiles. Larger cars from Horch and Mercedes were available for the higher command positions. Confiscated automobiles were also pressed into service, as was the case with trucks.

In order to conceal the need, flying Luftwaffe units were assigned civilian passenger cars which received a camouflage paint job right away.

War booty is examined at a Russian airbase; an Opel Blitz is along to cart off materiel.

Performance Data of Principle Vehicles Used by the Luftwaffe

	Opel Blitz	Borgward	Ford	Mercedes-Benz	MAN
Description	med. 3t truck	med.cross-country 3t truck	med. 3t truck	med. 3t truck	med. cross-country 4.5t truck
Type	3.6-36S	B 300 A/O	V 3000 S	L 3000 S	ML 4500 A
Years built	1937-1945	1939-1943	1941-1945	1940-1942	1940-1944
Wheels/drive	4 x 2	4 x 4	4 x 2	4 x 2	4 x 4
Engine	Opel 3.6 1	Borgward B 3000	Ford V 8	DB OM 65/4	MAN D 1040 G
Fuel	gasoline	gasoline	gasoline	diesel	diesel
Displacement	3226cc	3745cc	3922cc	4849cc	7980
Power	52 kW	57 kW	70 kW	55 kW	81 kW
Speed	85 km/h	80 km/h	85 km/h	70 km/h	63 km/h
Fuel supply	82 liters	120l	107l	90l	130l
Fuel consumption	25l/100km	29l/100km	30l/100km	20l/100km	25l/100km
Allowable total weight	5800kg	6440kg	5835kg	6790kg	10,000kg
Useful load	3300kg	3095kg	3300kg	3100kg	4950kg

	Mercedes-Benz	Mercedes-Benz	Henschel	Krupp
Description	heavy cross-country 4.5t truck	med. cross-country 3t truck	med. cross-country 3t truck	med. cross-country 3t truck
Type	L 4500 A	LG 3000	33 D1	L 3 H 163
Years built	1941-1944	1935-1938	1937-1942	1936-1938
Wheels/drive	4 x 4	6 x 4	6 x 4	6 x 4
Engine	DB OM 67/4	DB OM 67	Henschel Typ D	Krupp M 12
Fuel	diesel	diesel	gasoline	gasolinc
Displacement	7274cc	7413cc	10,857cc	7540cc
Power	82kW	70kW	74kW	81kW
Speed	66km/h	53km/h	53km/h	50km/h
Fuel supply	140l	112l	114l	150l
Fuel consumption	28l/100km	30l/100km	45l/100km	45l/100km
Allowable total weight	10,400kg	8500kg	9600kg	9300kg
Useful load	4658kg	2800kg	3500kg	350kg

This data is taken from the vehicles when outfitted as platform or flatbed trucks.

A French truck in front of an Fw 200 C belonging to KG 40, the rear gun position of which is covered with a tarpaulin to protect it against the weather.

Air transport truck of the Espenlaub firm with a single axle trailer, upon which a dismantled Arado training aircraft is being transported.

Left: A two-axle flatbed trailer with the fuselage of a crash-landed Bf 110 from ZG 26. A two-axle truck is towing.

Dismantling a Go 242 which had an off-airfield landing. One wing portin has already been loaded. A Hoanomag 100 horsepower tractor is being used to haul the trailer.

Towing Vehicles

Towing vehicles made up an important segment of the inventory of airfield vehicles. They were used for towing aircraft from parking positions to the airstrip when the aircraft could not taxi under their own power. Initially available for this purpose were trucks or heavy passenger vehicles as well as tractors ("Trecker"). Special towing vehicles, such as the heavy Faun ZR (110 kw power), the Hanomag SS 100 (74 kw), the light Hanomag *Schnellasttransporter* SS 55N (40 kw) and Lanz *Eilbulldog* Typ D 8531 (a 22 kw single cylinder engine), were towing vehicles which could be found in equal numbers in both the army and the Luftwaffe. Aside from the Faun tractor, which was derived directly from the manufacture of trucks, which in the west received a shortened truck chassis with moderate alterations to the drive train and transfer cases, the other vehicles were developed specifically for use as towing vehicles. They were well-suited for towing equipment on the roadways, but not so in terrain. Their military value was therefore limited, and aside form their use as towing vehicles for aircraft on airfields, they could be found in use as towing vehicles in industry, in the civilian sector and by the *Reichs-bahn* rail system. Nevertheless, the Hanomag S 100 LN as well as the Faun ZR and the ZRS railway model were produced up to the end of the war, and in some cases thereafter. In addition to these special-purpose machines, the Luftwaffe utilized any vehicle which had enough pulling power. In isolated cases, half-tracks were used, particularly for towing truck trailers. Even the use of animals for towing was not altogether rare, especially in 1944 when fuels were becoming more and more scarce.

In May of 1944 there was still no ground organization plan in support of turbojet and turbine aircraft on airfields, although the high fuel consumption during taxi of turbojet aircraft (Ar 234 and Me 262), caused in part by parking positions being particularly far from the runway due to security and camouflage considerations, demanded the creation of more powerful towing vehicles. These towing vehicles were to remain as simple as possible due to lack of raw materials. Additionally, the vehicles were made intentionally small in order to better camouflage them and make them easier to transport during deployments. The *Technisches Amt* examined vehicles already

An armored Renault-UE ammunition carrier, here with a 1000 kilogram bomb load on a wooden sled.

in mass-production which might satisfy these requisites. Aside from the planned utilization of the Fiat towing vehicle, which had already been proven in Africa, the test facility command introduced the creation of the *NSU-Kettenkrad (HK-101)*. In normal operations, this vehicle proved itself to be quite useful. Another vehicle utilized for flight operations was the "Scheuch-Schlepper," already in use in the Luftwaffe. This towing vehicle, manufactured by the Scheuch firm in Erfurt, had either a caterpillar or wheel drive. This could not be used in the case of the Me 163 rocket plane for understandable reasons, and so a special version with a towing apparatus was devised which pulled the aircraft aboard after landing in order to support the landing gear.

By using an integrated towing apparatus, which was developed by the testing facility, the aircraft, in particular the Me 262, could be safely towed at speeds up to 40km/h.

In get by with fewer overall towing vehicles and eventually be able to transport flight-ready aircraft on the Autobahn, a towing device was proposed by "Typenbegleiter Me 262" which would allow three to four aircraft to be towed on behind the other at the same time. Initial testing conducted in early 1945 was successful. All types of towing vehicles with pulling power of at least 200 kN were utilized.

Numerous small machines could be found on Luftwaffe airfields, like the torpedo cart and carts with starter engines, warm-up carts and, on airfields with sailplanes, towing winches. Additionally there existed a large number of trailers with the most varied of uses, for example wing transporters, ovens for baking as well as generator trailers.

In case of poor ground conditions, only a fully-tracked vehicle could be of help in bringing operational aircraft to the runway for takeoff or to parking positions. The ETC mounts on the fuselage of this Hs 120 B allowed it to accept eight 50kg bombs.

Towing Vehicle Performance Data

	Faun ZR	Hanomag SS 100	Hanomag SS 55N	Lanz D 8531	NSU HK-101
Description	wheeled tractor	wheeled tractor	prime mover	wheeled tractor	Kettenkrad
Years Built	1939-1946	1938-1943	1938-1943	1935-1941	1941-1944
Engine	KHD F6M517	Hanomag D 85	Hanomag D 52 SS 55	Lanz D 2539	Opel Olympia
Displacement	13540ccm	8553ccm	5195ccm	10,338ccm	1478ccm
Power	110 kW	74 kW	40 kW	42 kW	26 kW
Speed	60 km/h	45 km/h	33 km/h	30 km/h	70 km/h
Fuel supply	200-240 l	250 l	350 l	200 l	42 l
Fuel consumption	41l/100km	50l/100km	33l/100km	150l/100km	116l/100km
Allowable total weight	10,600 kg	8750 kg	4800 kg	4500 kg	1280 kg
Useful load	1050 kg	1150 kg	500 kg	–	325 kg
Pulling power	392.4 kN	196.2 kN	147.2 kN	–	4.4 kN

A captured caterpillar-track tractor with a two-axle trailer, which transported droppable munitions and ammunition.

A Fiat tractor tows an He 111
H-6 bomber in
Rostock-Marienehe.

Left: Another caterpillar
tractor with a home-made
canopy to protect the driver
from inclement weather.

The robust NSU Kettenkrad
tracked motorcycle in
particular served as a tower
for the Ar 234, as well as the
Me 262. It is shown here
during testing of snow skids
on the Ju 87.

Emergency Vehicles

In addition to the Opel "Blitz," the Phänomen "Granit" 25 H was used as a field ambulance, this aside from the recovery vehicles and converted panel trucks. The "Granit" was manufactured in large numbers by the Phänomen Werke AG located in Zittau/Sachsen beginning in 1930 as was a light commercial truck with rear wheel drive and an air-cooled internal combustion engine. As an ambulance it carried the nomenclature Kfz 31 and could accommodate four prone or eight sitting passengers. It could carry up to 1000 kilograms. Beginning in 1940 a newer version with a rounded radiator hood and driver's cabin was delivered as the "Granit 1500 S." Larger ambulances appeared on the chassis of the Büssing-NAG Typ 150, which was equipped with a choice of an internal combustion or diesel engine. Further, the Steyr 640, the Adler W 61 K as well as the Mercedes-Benz LE 1100 came into use, particularly with the army. In field operations it was very often the case that converted flatbed or panel trucks mounted on the chassis of the Ford Maultier and that of the 3-ton halftrack vehicle were used, especially during the muddy season in the East. Heavy passenger cars and busses, such as the Opel-Blitz "Wehrmachts-Omnibus," could accommodate seven prone and 16 sitting wounded passengers, or ten prone and six sitting wounded. In its configuration as a bus, it could accommodate 26 passengers.

The cross-country Magirus M 206 first came into use as a fire truck at the *Verkehrsfliegerschule* in the form of a tanker truck with a hose carrying trailer, and later on many other airfields. The first delivery of this three-axle vehicle by the C.D. Magirus firm in Ulm was in early January of 1934. By 1937, 529 chassis and 507 cross-country capable trucks had been assembled. They were equipped with the Magirus six-cylinder S 88 internal combustion engine.

The Magirus M 30 automotive tanker truck with its centrifugal pump and foam spray was considered very modern at the time of its delivery to the Lager-Lechfeld airport in 1934. Beginning in 1937, the Heinkel works in Oranienburg had the Magirus M 40 airfield tanker pumper truck. The Metz firm in Karlsruhe began to produce the TS 2.5 airfield pumper truck with a forward-mounted pump on the chassis of the Henschel 33 D1 beginning in 1937. By the end of 1941 the Magirus firm was also delivering this vehicle to the Luftwaffe.

The Magirus firm built the KS 15 fire truck on the Opel Blitz 3t chassis, which served in the Luftwaffe in numerous quantities. It was followed by the TS 1.5 tanker truck built on the basis of the Opel Blitz 3.6-6700 with all-wheel-drive.

The chassis of the Mercedes Benz L 4500 S medium truck also served the Luftwaffe as a fire fighting truck under the nomenclature KS 25. The TLF 25 tanker truck had an all-wheel-drive based on the Mercedes Benz L 4500 A.

There was a trailer for fire-fighting vehicles, the Ah. 301, which had a total weight of 1250 kilograms. Hoses and other necessary equipment was transported in this trailer. The hose tender truck (Kfz.309) could also assist in this mission of transporting hose. Occasionally this vehicle was not available at all.

The Mercedes-Benz six-cylinder L 3750 diesel engine capable of 74 kW power was often utilized on the Kfz. 309 hose tender truck. The firms Magirus in Ulm and Rosenbauer in Linz manufactured the superstructure for the Luftwaffe at the end of the 1930s. A trailer for hose transport (Ah.302) could be towed as well.

The Fw 58 "Weihe" (harrier) was utilized for the transport of sick and wounded personnel.

Above: Wounded are off-loaded from a Piaggio P.108 into waiting Opel ambulances.

Below: Fire fighting on a forward airfield using a foam extinguisher which could easily be used by small teams.

A crash-landed Do 17 — its engine caught fire and could not be saved due to lack of fire-fighting equipment.

Right: This is a vehicle belonging to the RLM fire brigade in Berlin, which fought fires on local airfields from time to time.

Pre-warming of the BMW 132 engines of this Ju52/3m belonging to TG/1 in the central sector of the Eastern Front.

A Hanomag vehicle tows a crash-landed Do 217 E-4 of III./KG 40 during its recovery at the end of 1940 in northern France.

Recovery of a Do 217 E-4 of I./KG 66 by using inflatable sacks to raise the aircraft so that it may be put on jackstands.

Left: A Bf 109 G-5s DB 605 engine is removed during overhaul.

Dismantling the engines of a Do 17Z by using a simple block and tackle. In the background is an Opel Blitz with a closed driver's cabin.

At almost all airfields, this type of block and tackle was used to change engines on Fw 190s and Ju 52/3m's.

A heavy gantry crane is used in changing engines (BMW 801ML's) on a Do 217 E-4.

This portable crane, transported on a trailer, proved its worth especially during DB 601 engine changes, in this photo on a Bf 110 *Zerstörer*.

At least three different
British cranes were captured
during the French Campaign
and remained in use by the
Luftwaffe. This model,
however, could not move on
its own and had to be towed.

Left: Carting off a crash-
landed Fw 200 C of I./KG 40
in western France.

Engine change using a
captured British crane. This
He 111 belonged to KG 100.

Aircraft Fuel Tanker Trucks

For heavy aircraft fuel trucks the commercially available chassis of the Büssing VG and the Daimler Benz L5, each with more than 5 tons capacity were utilized in the early 1930s. Another commercial product was the medium-weight fuel tanker truck mounted on the Daag-Lkw ACO chassis, which ad up to four tons capacity. A further aircraft fuel tanker was mounted on the Mercedes LG 3000 (F.K.Kw, Kfz.384). The capacity of this tank was 3,500 liters. These vehicles commonly had a forward-mounted exhaust and an altered driver's cabin with a foldable top. Beginning in 1937 the Luftwaffe utilized the Opel "Blitz" (Kfz.385) with a tank superstructure as a tanker truck capable of 3 tons capacity in increasing numbers. Some of the Opel "Blitz" trucks had an open driver's compartment, modified especially for airfield use. The Luftwaffe used these vehicle to transport other fuels such as J2 as well and C and T fuels (C-Stoff and T-Stoff). In these cases it was prudent not to change fuels transported until after the tank interior had been thoroughly cleaned out. The reaction, in the case of mixing the smallest amount of T-Stoff with C-Stoff, for example, was disastrous. Uncommon was the Ford Lkw V 3000 S which was a tractor-trailer configuration with a mounted tank unit. This rare vehicle is known to have been used on deployment airfields during the Battle of Britain and in 1943 on an airfield in southern Russia. further, the Krupp L 3 H 163 chassis offered the opportunity for the mounting of a fuel tank with a capacity of 3600 liters. The fuel trucks predominantly had an engine driven pump for fuel transfer as well as a supplemental fire extinguisher.

Very often on airfields there were flatbed trucks with gasoline canisters fastened to them, or a fuel tank on the back with hand-or motor-driven pumps.

Additionally, there was a fuel tanker truck used as a trailer (Ah.454) which had a capacity of 2625 kilograms and a total weight of 6825 kilograms, which was moved by special towing vehicles, for example half-tracked or wheeled tractors.

Because the pump system of this captured tanker truck was defective, a trailer-mounted pump was enlisted for the task of filling up this Ju 88 A-4 of KG 30.

In the early years of the war, single-engine fighters were usually refueled directly from fuel tanks mounted on trucks by using a hand pump.

A number of the commercial Opel Blitz tanker trucks were taken over by the Luftwaffe at the beginning of the war.

A tanker vehicle mounted on an Opel chassis in use on the eastern front. In front is an Ah.454 aircraft fuel tanker-trailer.

A Kfz.384 aircraft fuel truck which had the medium cross-country capable truck chassis.

Another Kfz.384 with II./KzbV 1, which was later II./TG 1 in 1943 on the southern sector of the eastern front.

The Renault-UE tractor was quite capable of pulling the two-axle Ah.454 trailer.

A light wheeled tractor used in towing fuel tanks was quite sufficient under general field conditions.

Oil Transporters

The oil transporters (Fl 65851) were set-up as single-axle trailers with two roadwheels. Two adjustable supports (forward and on the rear of the chassis) ensured a secure footing. The shaft's angle was adjustable by means of two hinges. The oil tank's inclination could be adjusted independent of the relative height of the coupling on the vehicle pulling it. On smaller airfields it was often the case that the maintenance crew pushed the trailer themselves.

Later, simpler versions had disk wheels with rubber tires and only one shaft.

The oil tanks were equipped with one hand pump and one electric pump. The electric pump worked only in conjunction with the towing vehicle (electricity being supplied by running its engine) or with a generator. The cart had separate chambers so that various lubricants as well as coolants could be transported in one vehicle. Those hoses necessary for servicing an aircraft could be laid around the trailer while under way. In addition to this trailer tank vehicle there were initially hand carts for lubricants and coolants. These were fitted with two spoked wheels and a smaller steerable support wheel in front. A metal handle aided in manually pulling or positioning the cart. It was equipped with a hand pumping system and a heater which helped keep the oil in a more liquid state during cold weather.

The use of this tank cart was often only seen on the larger airfields with a hard-surfaced runway. On forward airfield the maintenance crews primarily used oil barrels for filling the lubricant containers. These barrels were sometimes mounted on trucks.

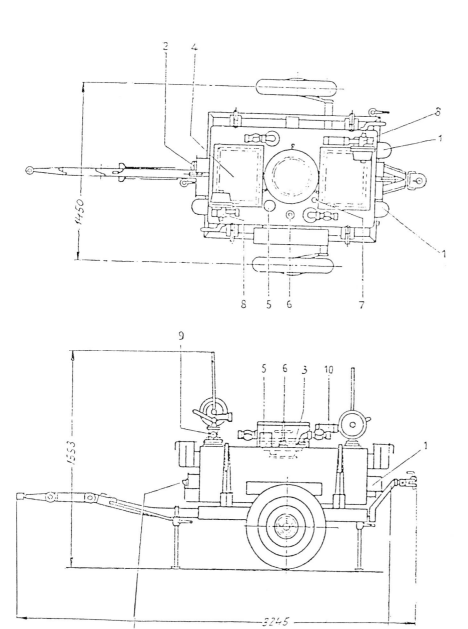

Öltank- und Kühlmittel-
wärmewagen

1. Dampfein- und -austrittsöffnung
2. Anschluß für elektrische Beheizung
3. Einfüllöffnung
4. Reinigungsöffnung
5. Thermometer
6. Inhaltsanzeiger
7. Dampfpfeife
8. Allweiler Flügelpumpen
9. Flanschfilter
10. Handhabefilter
11. Kupplungshalter

Start Carts

Start carts or generator vehicles had the mission of relieving the strain on the aircraft batteries and to supply sufficient energy during equipment and weapons tests as well as during engine start-ups. Additionally, those batteries located in the lower fuselage were charged up. For this purpose, a generator was mounted on a two-wheeled trailer and was driven by a two-cycle engine.

The vehicle body was mounted on a steel-tube chassis which was initially fitted with disk wheels, but later received spoked wheels with rubber or steel tires/wheels. In the middle of the start cart there was a cast iron plate with all the switching components. The engine was installed behind this. A hood, which was folded back while the generator was in operation in order to keep the engine from overheating, protected the engine from the influences of weather. Power was carried to the external power socket of the aircraft via a cable with a special plug. When not in use, and when the vehicle was in tow, this cable lay in an open box in front of the main switch.

Fueling a Ju 88 A-4 (3Z+CH) of 1./KG 77 using an Italian tanker in southern Italy. A start cart can be seen in the background. An experimental model of the Fw 191 during factory testing by Focke-Wulf. Fueling is done from an old tanker trailer.

Using a hot-air warming device to warm up the engines of an He 111 H-6.

Winter flight operations in 1940 in the West. Neither the camouflage paint nor the little pine trees can contribute to the camouflaging of this Bf 109E.

Not all ground crews in the Russian campaign were equipped with the appropriate clothing for the harsh weather conditions. In the foreground is a heater.

Because aircraft repair halls were almost completely unavailable in the East, heated, collapsible maintenance huts were mass-produced.

Maintenance cabin of Transportgeschader 1, which permitted work to be done on the external engines of this Ju 52/3m in extreme cold.

The runway is cleared by a Model E snowplow, which could clear a path 3.5 meters wide. Noteworthy are the drop tubes and fuselage-mounted machine gun turret on the parked Bf 110.

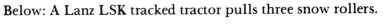

Below: A Lanz LSK tracked tractor pulls three snow rollers.

This Fiat caterpillar tractor was used to pack the snow at the airfield of this *Zerstörer-Ergänzungsstaffel.*

An operational aircraft of **KG 77** during maintenance in Italy. In the foreground is an oilcart with a hand pump. This type of trailer had huge wooden wheels.

In case there were no snow rollers available, large wooden constructions were often used which, as seen here in 1943 in Russia, were pulled by a team of horses.

The Citroen P 19 half-track vehicle was used on many Luftwaffe airfields in the West, for example that of KG 54.

Start carts helped spare the batteries and could be seen everywhere. This photo was taken at the Henschel factory airfield near Berlin.

Below: A Ju 88 A-4 of **KG** 3 during maintenance and rearming operations under sunny winter conditions in Russia, 1943.

Modern trailers were used to protect the needed oxygen and enabled the aircraft oxygen systems to be refilled directly (France, 1940).

Jet Engine Aircraft Transporters

A transport trailer was built especially for carrying jet turbine engines which had a load capacity of 1.5 tons and underwent a short test period. The vehicle also came in handy when assembling jet engines and could be utilized as an engine test stand as well. Vehicles of all sorts served to tow these trailers.

The Me 262 engines were then hung under the wings of the aircraft via a crane. This was made easier by mounting of heated fastener rings which grasped the leading edge of the wing. A special rolling harness which could be broken down into eight parts was used in the case of the Ar 234.

An interesting towing arrangement with an He 162 of I./JG 1 and a generator trailer in 1944 in northern Germany.

Transporting an HeS11 jet engine at the Heinkel factory in Stuttgart.

Jettisonable Ordnance & Equipment

The possibility of tossing explosive ordnance from an aircraft received quite a bit of interest, even before the First World War. At first these weapons were limited to handgrenades, which were occasionally wrapped together for increased effect. This was followed by 4.5 kilogram carbonite bombs which had an iron rod with impact disks which prevented the weapon from penetrating below the ground's surface and maximum explosive effect on the surface was achieved. In the Second World War the Luftwaffe again used this type of bombing. Proximity fuses, so called "Dinorstäbe" permitted the bombs to explode above the surface and achieved a fragmentation effect. Other jettisonable munitions utilized in the First World War and dropped by mine layers such as the "Priesterminen" and the "Fliegermaüse" mines, 0.5 kilograms charges of explosive which were packed four at a time into ejector tubes. In the course of the war bombs and other jettisonable munitions were used with increasing accuracy and bombs were developed with greater and greater explosive effect.

As a result of the Versailles Treaty, further research had to come to a halt in the 1920s, and only after rearming period had begun was an attempt made to create more effective jettisonable. By the beginning of the Second World War, the Luftwaffe had at its disposal the 1kg *Elektron-Stabbrandbombe* incendiary bomb introduced in 1933, the SD 10 fragmentary bomb as well as the SC 50 and SC 250, and beginning in 1935 the SC 500. One year later, the 500kg LMA and 1000kg LMB landmines were introduced.

Due to a lack of heavy jettisonable munitions, the Luftwaffe introduced the SC 1000 in 1940 and on 29 July 1941 succeeded in dropping an SC 1800 on the Citadel in Brest-Litovsk.

The SC 500 was capable of penetrating an armor plate of 40mm thickness. The average destruct radius was about 26 meters, the effective fragmentation radius was 190 meters. The SC 500 could tear a crater almost 10 meters deep and 16 meters wide in earth of normal density.

In addition to these explosives, which were used against standard targets, a series of anti-armor bombs were developed beginning in 1938 (PC 500 to PC 1600), which were intended to be primarily utilized against hardened targets or heavily armored warships. The had a delayed fuse so that the detonation occurred after the weapon had penetrated the armor. The penetrating power of

Jettisonable munitions lay waiting for use. At the tip of the bombs can be seen hanger eyelets, used for loading in the vertical magazine. There are four carrying handles on each bomb to make transport somewhat easier.

the PC 1000 was about 100mm of armor or 1.5 of reinforced concrete. These values increased in the case of the PCz1400 by about 20 percent. During the war, the anti-armor bombs PD 500 and PD 1000 were also available.

Bombs such as the PC 500 RS, PC 1000 RS and PC 1800 were manufacture in limited numbers with propulsion units, but were eventually replaced by remotely guided jettisonable munitions such as the Hs 293 A-1/A-2 or the PC 1400X "Fritz X."

There were also high-explosive bombs with enhanced fragmentation (thick-walled high-explosive bombs), the SD-series (SD 1 to SD 1700).

The introduction of the SDz1 took place in 1942 and coincided with shaped charge SD 4 HL bomb which, when utilized by fighter-bombers, proved effective against vehicles. The lighter models in the SD-series, up to SD 9, were kept in air-drop containers. (AB). An AB 250 could be filled with either 224 SD 1's, 144 SD 2's or 17 SD 4's. Incendiary bomb models B1, 2, 4 and 10 were also transported in this manner. An AB 250 could disperse thirteen B 10's and six B 1's from 1000 meters over a 30 meter by 50 meter area. The most common versions were the AB 70, AB 250, AB 500 and AB 1000.

Two BSB 700/XII incendiary bomb hoppers were necessary for mass-drop bombings of the B1 incendiary stick bombs, which were hung on the wing hard points. Each hopper could hold about 700 of these small bombs.

On the Ju 87, four bundles of five SD 10 bombs could be observed under the wing hard points. The bundling was accomplished by means of a protective covering.

Deployment conditions on dislocated airfields were just as varied as the vehicles found upon them. In the background is a captured tanker truck.

A Ju 88 A-5 is made ready for takeoff on its next mission.

Left: The loading rod helped slide the SC-50 bombs in to the vertical magazine of the He 111.

The most commonly used jettisonable munitions of the Luftwaffe (left to right): SC 50, SC 250, PC 500, SC 500, SC 1000 and the SC 1800.

In 1944, new universal jettisonable containers were introduced which could serve as 500 liter drop tanks, carry 500 kilograms of fragmentary bombs, be utilized as jettisonable supply containers or be utilized to carry personal gear during deployments.

In the case of the Fw 190 A there was, at the end of 1942, a mounting grate (ER 4) for four 50kg bombs because there were not enough Fw 190 F fighter-bombers with under-wing hard points available. For the Fw 190 E/F a saddle mounting for four 50kg bombs was utilized. Added to the inventory of small incendiary bombs in November of 1943 were the B C 50, B C 250 and the B C 500. The B C 500 could spread its incendiary contents over a 14,300 cubic meter area and was intended for use in igniting flat target areas.

Beginning in 1942, several SB 2500 bombs were available on the Front. The bombs had a length of 3.90 meters and only in use for a short period due to lack of suitable flight equipment. There also existed a few versions with smaller dimensions, which were often carried by the Do 217 E-4 and K-1 as well as the He 177 A-3 and A-5 aircraft. Later the SB 2500 casings were filled with the so-called "Triamen-Mischung 105" mixture in order to be able to successfully engage commercial vessels from close quarters out to 70 meters from the hull. The explosive effect with this mixture was so strong that when the bomb was detonated above ground, it could be expected that buildings within a 1000 meter area would be destroyed and glass would be broken in the surrounding area.

There were other illuminating bombs (BL C), bombs for war materiel (KC), air mines (SB) and maritime bombs of various types and sizes. There were also captured bombs, such as Russian 100kg bombs and French 20kg explosive bombs which were utilized in particular by night-fighter units on the Eastern Front. The chart to the right gives an overview of those bombs most frequently used by the Luftwaffe.

Bombs up to 500 kilograms in weight were primarily transported to the airfields in crates, the heavier bombs in transport racks. Transportation on the airfield itself was accomplished by means of simple cages by hand or on sleds in the Winter. Various three-wheeled bomb carts were used by the ground crews to hang the bombs in their racks aboard the Ju 87 and all fighter-bombers (Bf 109, Fw 190). On most combat aircraft the bombs could be loaded by a special harness.

Type	Gross Weight	% of Explosive	Length Diameter (cm)
SC 10	10	42	58.5	8.6
SC 50	2 50	50	109.5	20.0
SC 250	250	54	165.1	36.8
SC 500	500	54	202.2	47.0
SC 1000	1000	53	280.0	65.4
SC 1800	1800	56	350.0	66.0
SC 2000	2000	60	350.0	66.0
SC 2500	2450	69	389.5	82.9
SD 1	1	20	35.0	5.0
SD 2	2	11	30.3	7.8
SD 4/HL	4	9	31.5	9.0
SD 9/HL	9	9	31.0	12.0
SD 50	50	32	109.0	20.0
SD 70	70	30	128.0	20.0
SD 250	250	32	193.8	36.8
SD 500	500	28	200.7	14.2
SD 500A	500	33	202.2	44.7
SD 500E	500	22	174.0	39.6
SD 1400	1400	23	284.0	56.3
SD 1700	1700	41	330.0	66.0
PC 500	500	20	200.7	29.5
PC 1000	1000	16	210.0	50.0
PC 1400	1400	14	283.6	56.2
PC 1600	1600	18	318.5	66.0
PD 500	500	6	210.0	27.6
SB 1000	1000	85	280.0	65.4
SB 1800	1800	83	350.0	66.0
B 1	1.1	18	35.0	5.0
B 2	2.2	23	59.5	5.0
B 4	4.2	29	70.0	8.0
B 10	11	32	107.0	11.5
B C50	45	43	109.5	20.0
B C250	185	43	193.8	26.8
B C500	315	50	200.7	47.0

Right: Here, a passenger car belonging to JG 51 waits for a newly decorated Knights' Cross recipient to arrive.

Hydraulic bomb loader, which was available at all deployed units and could be used to lift loads of all sorts.

Below: Loading a Ju 87-D ''Stuka'' under winter conditions.

By using the bomb loading cart, almost all heavy loads could be placed in the proper position directly under the fuselage of the He 111, He 177 or Ju 88 aircraft.

Listing of Some of the Luftwaffe's Most Commonly Used Wehrmacht Vehicles

Utilization	Commonly used chassis
crew vehicle	Krupp L 2 H 143, heavy passenger car
med. weather	medium cross-country truck with closed cabin:
printing truck	
radio relay truck	Henschel 33 D 1, Büssing-NAG 3 GL 6
teletype truck	Krupp L 3 H 63/163, Magirus 33 G 1
teletype relay	
radio power supply	
radio truck	
crane	Borgward B 3000
rotating crane	cross-country truck of the 4.5 ton class with support, towing equipment
	MAN ML 4500, Büssing-NAG 4500 Mercedes-Benz L 4500
heavy tractor	Faun ZR, Hanomag ST 100
light tractor	Hanomag Schellasttrsp. SS 55N
light spotlight	Krupp-Protze L 2 H 143,
truck	heavy unit-level auto Horch, Ford
passenger car	Adler V 40 T, medium cross-country (air transportable) passenger car
passenger car	all Wehrmacht passenger cars

Applying a block and tackle to lift an SC 1800 under the fuselage of a Ju 88 A-4.

Below: Loading a Ju 188 A-2 of KG 6 with an SC 2500 "Max" by using a block and tackle.

Transport via bomb carts or sleds, in accordance with the size of the bomb. Below is one of the heaviest, the SC 1800.